Grades K-4

Focus On Elementary Physics

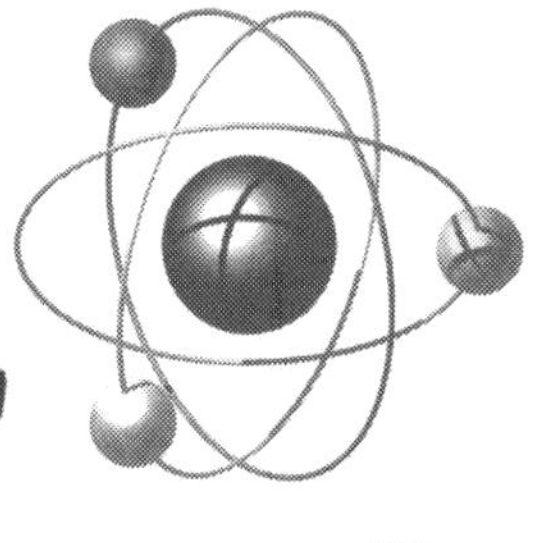

Teacher's Manual

Rebecca W. Keller, PhD

Real Science 4 Kids

Illustrations: Janet Moneymaker and Rebecca W. Keller, PhD

Focus On Elementary Physics Teacher's Manual
ISBN 978-1-936114-64-1

Published by Gravitas Publications, Inc.
www.gravitaspublications.com

Printed in United States

A Note From the Author

This curriculum is designed to provide an introduction to physics for students in kindergarten through fourth grade. The *Focus On Elementary Physics Laboratory Workbook* is intended to be the first step in developing a framework for real science. This *Teacher's Manual* will help you guide students through the laboratory experiments as they develop the skills needed for the first step in the scientific method—making good observations.

There are five sections in each chapter of the *Laboratory Workbook*. Section I, *Observe It,* helps students explore how to collect data and make good observations. The second section, *Think About It,* provides questions to help students think about what they observed. Section III, *What Did You Discover?,* gives students an opportunity to summarize the observations they have made. The section called *Why?* provides a short explanation of what students may or may not have observed. Finally, there is an extra activity for students to experiment with in Section V, *Just For Fun.*

Each experiment takes up to 1 hour to do. The materials needed for each are listed in the charts on the next two pages as well as at the beginning of each experiment.

Enjoy!

Rebecca W. Keller, PhD

Materials at a Glance

Experiment 1	Experiment 2	Experiment 3	Experiment 4	Experiment 5
2 tennis balls Other objects such as: apple orange rubber ball cotton ball feather	marshmallow rubber ball tennis ball lemon or lime rock banana other objects pliers	toy car stiff cardboard or wood board about .3 meter (1 ft) wide and 1 meter (3 ft) long marshmallows, several Optional: pennies tape	1 small glass marble 1 large glass marble	3 large lemons knife 3 copper pennies older than 1982 3 galvanized (zinc coated) nails LED (Radio Shack #276-30700 or similar LED) 4 pairs alligator clips plastic coated copper wire wire clippers small Phillips screwdriver Optional: duct tape

Experiment 6	Experiment 7	Experiment 8	Experiment 9	Experiment 10
2–3 rubber balloons string or thread, at least 2 meters (6 ft) cut in half scissors Materials such as: cotton clothing wool clothing silk clothing plaster wall metal surface leather surface wooden surface	lemon battery (see Exper. 5 materials list) glass of water table salt, 15 ml (1 Tbsp.) Test materials: Styrofoam plastic block cotton ball nickel coin metal paper clip plastic paper clip	2 bar magnets with poles labeled "N" and "S"	2 plastic or glass prisms flashlight colored pencils or crayons Suggested: colored light bulb halogen light bulb fluorescent light bulb mercury light bulb Optional: cardboard box	2 marbles 3 playing cards shallow jar top vinegar baking soda measuring spoons Suggested: dominoes blocks electric car electric train marshmallow tongue depressor steel ball

Materials at a Glance

By type and quantity

Equipment	Materials	Materials (Cont.)
alligator clips (4 pairs) car, toy flashlight glass, drinking (1) knife magnets, bar, with poles labeled "N" and "S" (2) pliers prisms, plastic or glass (2) scissors screwdriver, small Phillips wire clippers	balloons, rubber (2-3) block, toy, plastic (1) cardboard (stiff) or wooden board, about .3 meter (1 ft) wide and 1 meter (3 ft) long cotton ball (several) crayons or colored pencils jar top, shallow (1) LED (Radio Shack #276-30700 or similar) (1) marble, small glass (1) marble, large glass (1) marbles, any (2) materials such as: cotton clothing leather surface metal surface plaster wall silk clothing wooden surface wool clothing measuring spoons nails, galvanized (zinc coated) (3) nickel coin (1) objects to drop (several) objects to squeeze (several) paper clip, metal (1) paper clip, plastic (1) pencils (colored) or crayons pennies (3 copper, older than 1982) playing cards (3) rock (1) rubber ball (1) salt, table, [15 ml (1 Tbsp.)] string or thread, at least 2 meters (6 ft) cut in half Styrofoam (small piece) tennis balls (2) wire, plastic coated copper	Optional: box, cardboard (1) duct tape pennies (several, any) tape Suggested: ball, steel blocks, toy (several) car, toy electric dominoes (several) feather (1) light bulb, colored light bulb, fluorescent light bulb, halogen light bulb, mercury tongue depressor train, toy electric

Food Items
baking soda [about 30-60 ml (2-4 Tbsp.)] banana (1) lemon (1 any size) lemons (6 large) marshmallows, several vinegar [about 30-90 ml (2-6 Tbsp.)] Suggested: apple (1) lime (1) orange (1)

Contents

Experiment 1

Falling Objects

Materials needed

- 2 tennis balls
- other objects such as:
  apple
  orange
  rubber ball
  cotton ball
  feather

Objectives

In this unit students will try to determine if Galileo was right.

The objectives of this lesson are:

- *To help students compare their observations with a scientific discovery.*
- *To help students compare different observations.*

Experiment

I. Observe It

In this section students will observe how two objects fall when they are released at the same time.

❶ Have the students hold two tennis balls in their hands at chest level.

❷ Have the students release the two tennis balls at the same time.

❸ Help them observe how the objects land on the ground. Ask them the following questions.

- *Did the objects land at the same time?*
- *Is one object heavier or lighter than the other object?*
- *Do you think it matters how high you hold the objects? Why or why not?*
- *Do you think the shape of the object matters? Why or why not?*

❹ Help the students record their observations in the *Observe It* section of their *Laboratory Workbook* (see example on next page).

❺ Have the students repeat the experiment using different combinations of objects. Have them compare at least four different pairs of objects. For each set of objects, help the students record their observations.

In the spaces next to **Object 1** and **Object 2,** have the students write the names of the objects they are using.

Then have them draw or write a description of what they see. Help them make good observations by asking questions such as:

- *How heavy does Object 1 feel in your hand?*
- *How heavy does Object 2 feel in your hand?*
- *Does Object 1 feel heavier or lighter in your hand than Object 2?*
- *Is it easy to release both objects at the same time? Why or why not?*
- *Describe the shape of Object 1.*
- *Describe the shape of Object 2.*

Object 1 *apple*

Object 2 *tennis ball*

(Answers may vary.)

The apple feels heavier in my hand than the tennis ball.

When I drop the objects, I have a hard time seeing which one lands first. They look like they land together, but I am not sure.

I can release the objects at the same time from my hands.

The apple is shaped differently than the tennis ball, but they still land at the same time.

II. Think About It

❶ Help the students think about their experiment and make any observations about how easy or difficult it was to perform the experiment.

- *Could they see the objects fall to the floor?*
- *Could they determine if both objects fall to the floor at the same time?*
- *Was it easy or difficult to release the objects from both hands at the same time?*

❷ Help the students think about ways to vary their experiment.

- *If they can't see the objects fall to the floor, maybe they can get a parent, sibling, or friend to make the observations.*
- *What would happen if the objects were released from a greater height? Students could stand on a sturdy chair or bench to see if height makes a difference.*
- *What would happen if the objects were released from a lesser height? Students could release the objects at their waist or at their knees.*

❸ Help the students repeat the experiment for one of their object pairs. Have them vary only one parameter at a time for one set of objects. For example, they may want to hold the objects higher and have a friend observe how they fall. But this is changing two parameters (the height and the observer). Explain to them that scientists try to vary only one parameter at a time so that they can make comparisons to previous experiments, noting what difference, if any, changing one parameter had on the experimental results. If more than one parameter is changed at the same time, it will be unclear which change caused different results.

❹ Help the students record their observations.

III. What Did You Discover?

The questions can be answered verbally or in writing, depending on the writing ability of the student. With these questions help the students think about their observations. There are no "right" answers to these questions, and it is important for the students to write or discuss what they actually observed.

IV. Why?

Help the students understand that two objects of different weights will fall to Earth at the same speed. Both objects have the same amount of gravity pulling on them. So, both objects start with the same force. Since the speed of an object is independent of its weight and since both objects have the same force, they will both fall to Earth at the same speed.

V. Just For Fun

The students can try an experiment with an object that will be much lighter than the objects they have been using. A cotton ball or a feather would work. They will discover that if an object is too light, it will float to the ground and not fall at the same time as a heavier object. Explain to them that if the two objects are put in a vacuum, they will fall at the same time. Even an apple and a feather or cotton ball will fall at the same time. In the vacuum there is no air resistance. Outside the vacuum, the air pushes up on the cotton ball, and because the cotton ball is light enough, the air will slow it down.

Experiment 2

Get To Work!

Materials needed

- marshmallow
- rubber ball
- tennis ball
- lemon or lime
- rock
- banana
- other objects
- pliers

Objectives

In this unit students will explore the concept that work happens when energy is used to create a force that moves an object, changes its shape, or changes its velocity.

The objectives of this lesson are:

- *To help students explore the concepts of work, force, and energy.*
- *To help students make careful observations.*

Experiment

I. Observe It

In this section students will compare what happens to different objects when they apply a force to them with their hands.

❶ Have the students observe a marshmallow. Have them write or draw a description of the size, shape, and color of the marshmallow (see example on following page).

❷-❹ Have the students hold the marshmallow in one hand and then press the marshmallow with the palm and fingers of their hand. Use the following questions to help the students observe how much effort they used while squeezing the marshmallow.

- *Was it easy or difficult to squeeze the marshmallow?*
- *Were you able to squeeze the marshmallow completely? Why or why not?*
- *Would you describe the marshmallow as "hard" or "soft?"*

❺ Help the students record their observations in the "Observe it" section of their workbooks.

❻ Have the students repeat this exercise with several other objects.

(Answers may vary.)

Marshmallow

Before

white
cylinder
light
fluffy

After

sticky
smashed
white

II. Think About It

1. Help the students think about their experiment and make any observations about how easy or difficult it was to perform.

- *How did the objects feel in their hands? Were some objects heavy? Some light?*
- *Have the students discuss the size, shape and color of the different objects. Were some difficult to hold? Did all the objects fit in their hands? Were some too large to hold?*

2. Have the students discuss how difficult or easy it was to squeeze the different objects.

3. Help the students create a summary of their observations. The two columns in the *Laboratory Workbook*—**Easy to Squeeze** and **Hard to Squeeze**—are very general categories but will give the students a way to sort the different objects.

4. Have the students review Chapter 2 and the summary statements in the *Student Textbook*. Have them recall that work is what happens when a force moves an object, changes its shape, or changes how fast or slowly an object is moving. Help them make the following connections:

- *Their hand is using energy (from their body) to squeeze the object.*
- *Their hand is generating a force on the object (the squeezing).*
- *The object is either changing shape as the force is applied or is not changing shape.*
- *The amount of work done is related to how much the object changes shape.*
- *The more the object changes shape, the more work is done.*

Help the students determine when the most work was done by squeezing the objects with their hands. Without telling them the answer, again remind them that the amount of work that is done depends on the amount the object changes shape. This means that even if they squeeze an object with all their effort, using lots of energy and applying lots of force, if the object does not change shape, they do no work.

Have the students circle the object they believe had the most work done to it when it was squeezed.

❺ Have the students put a rectangle around the object they believe had the least work done to it when it was squeezed.

III. What Did You Discover?

The questions can be answered verbally or in writing, depending on the writing ability of the student. With these questions, help the students think about their observations. There are no "right" answers to these questions, and it is important for the students to write down or discuss what they actually observed. Help them explore how the answer they got may be different from what they thought might happen. If they guessed that more force would equal more work, they may be surprised to discover that this is not necessarily true.

IV. Why?

Discuss with the students how their body can be used as a tool to measure force and energy. Because their hands have nerve endings, they can sense the hardness or softness of different objects. Also, they can observe their muscles and breath as they squeeze different objects. This gives them an idea of how much energy is needed to generate different forces. The harder they squeeze an object, the more force they are generating.

Discuss with the students that the use of more force and more energy does not necessarily mean they did more work. The shape of an object has to change before work is done. If a force is applied, say to a rock, but the shape does not change, no work is done. Discuss how it may feel as if they did work, but without an observable outcome, no work was done.

V. Just For Fun

Without a pair of pliers, it would be difficull for most young students to squeeze a tennis ball. However, with a pair of pliers it becomes much easier. The pliers are a form of tool called a lever, and they actually multiply the amount of force exerted on the tennis ball. That is, with a little applied force on one end (the handle) there is much greater force applied at the other end (the tennis ball).

Experiment 3

Moving Energy in a Toy Car

Materials needed

- toy car
- stiff cardboard or board, approximately .3 meter (1 ft) wide and 1 meter (3 ft) long
- several marshmallows

Optional

- pennies
- tape

Objectives

In this experiment students will explore converting gravitational stored energy into kinetic energy.

The objectives of this lesson are:

- *To help students observe how one form of energy is converted to another form of energy.*
- *To help students learn how to collect data and create a table of their results.*

Experiment

I. Observe It

In this section students will perform a simple experiment to explore gravitational potential energy and see how a toy car will have more kinetic energy (moving energy) as it gains more gravitational potential energy. The students will need a toy car and either a stiff cardboard sheet or a board at least 1 meter (3 feet) in length.

❶ Have the students place the board or cardboard sheet on the ground with the toy car at one end.

❷ Even though the car is not moving, have them write or draw their observations. You can explain to the students that this is the "starting point" in the experiment. Even though nothing is happening, scientists always record their observations at a starting, or reference, point. A starting or reference point gives scientists an observation with which to compare other observations.

Car—Flat

(Answers may vary.)

Car is not moving.

It is just sitting still

on the ramp.

❸ Have the students stand at the end of the cardboard sheet or board where the toy car is placed. Make sure the toy car is facing toward the other end of the cardboard sheet or board. Have the students lift the cardboard sheet or board to the level of their ankles. The toy car may or may not move at this height.

❹ Have the students record their observations. Guide their observations with the following questions:

- *Is the toy car higher or lower now that you have lifted the board to your ankles?*
- *Did the toy car move? If so how far?*
- *If the toy car moved, how fast or slow did it move?*
- *If the toy car moved, did it make it all the way to other end of the cardboard sheet (or board)?*

Car Lifted to Ankles

(Answers may vary.)

Car moved slightly

down the ramp. Only

an inch or so. It did

not move very fast.

❺-❿ Repeat steps ❸-❹ for the car lifted to the knees, then the hips, and then the chest.

Collect Your Results

Have the students complete the table provided by entering the results they recorded on previous pages.

II. Think About It

❶ Help the students think about their experiment and make any observations about the toy car. You can use the following questions to guide discussion:

- *Did the toy car move easily once the board was lifted?*
- *Were the wheels sticky, or did they move easily?*
- *Do you think the car was heavy enough to move down the board?*
- *Was the car made of plastic or metal?*
- *Do you think what the car was made of could make a difference in the results?*
- *Was the cardboard sheet (or board) smooth or rough?*
- *Do you think that what the ramp was made of could make a difference in your experiment?*

❷ Review with the students the last part of Section 3.2 of the *Student Textbook* which covers gravitational stored energy. Any object that is elevated from the ground has some amount of gravitational stored energy. When the object falls to the ground, the gravitational stored energy is converted to kinetic energy. Objects that are higher than other objects have more gravitational stored energy.

❸ Have the students guess which car height had the least amount of gravitational stored energy and which height had the greatest amount of gravitational stored energy.

❹ The car with the least amount of gravitational stored energy was the car that was sitting flat on the ground. It had zero stored energy because it was in contact with the ground. Have the students put a circle around this item.

❺ The car with the greatest amount of gravitational stored energy was the car that was lifted to the chest. Have the students draw a rectangle around this item.

Car Flat

Car Lifted to Ankles

Car Lifted to Knees

Car Lifted to Hips

Car Lifted to Chest

III. What Did You Discover?

The questions can be answered verbally or in writing, depending on the writing ability of the student. With these questions help the students think about their observations. There are no "right" answers to these questions, and it is important for the students to write or discuss what they actually observed. Help them explore how the answers they got may be different from what they thought might happen. Help the students understand the concept of converting gravitational stored energy to kinetic energy. The more gravitational stored energy the car has, the more kinetic energy it will generate as it moves down the ramp.

IV. Why?

Discuss with the students how the toy car was able to move once it was lifted from the ground. Help the students understand that as their body lifted the ramp, the car was lifted and "gained" gravitational stored energy. Help the students see that as they lifted the car higher, they added more gravitational stored energy to the toy car.

V. Just For Fun

Have the students observe how kinetic energy can do work. As they raise the ramp and the toy car moves, the toy car's gravitational stored energy is converted to kinetic energy. Help the students qualitatively determine how much kinetic energy is needed to smash a marshmallow. They may have to lift the car and ramp higher than their heads. Also, it is possible that no matter how high they lift the ramp, the marshmallow won't smash. If this happens, you can ask them what they think might happen if they used a heavier car or taped some pennies to the car they are using to make it heavier.

Experiment 4

Rolling Marbles

Materials needed

- 1 small glass marble
- 1 large glass marble

Objectives

In this experiment students will explore inertia, mass, and friction.

The objectives of this lesson are:

- *To help students observe how mass and inertia are related and how the force of friction slows and eventually stops kinetic energy.*
- *To help students learn how to collect data and create a table of their results.*

Experiment

I. Observe It

In this section students will perform a simple experiment to explore inertia, mass, and friction. The students will use two types of marbles—a small glass marble and a large glass marble (a marble that is several times larger than the small marble).

❶ Have the students take the small glass marble and the large glass marble and roll them on a smooth surface.

❷ Have the students observe how each marble rolls. Guide their inquiry with the following questions:

- *Do the marbles go straight?*
- *How far do the marbles go?*
- *Where do the marbles stop?*
- *How do the marbles stop?*

❸ Have the students record their observations.

Small and Large Marbles Rolling on Smooth Surface

(Answers may vary.)

The small marble rolled across the wood floor from the chair to the couch. It stopped when it hit the couch, and it bounced back and stopped a few inches from the couch.

Have the students draw what they observed.

❹ Have the students take the same marbles and roll them across a carpeted surface or a grass lawn. This "rough" surface has more friction than a smooth surface.

❺ Again help the students think about their observations. Guide their inquiry with the following questions:

- *Do the marbles roll straight?*
- *How far do the marbles go?*
- *Does one marble go farther than the other?*
- *How do the marbles stop?*
- *Is one marble harder to roll than the other?*

❻ In the space provided, have the students record their observations. Have them pay particular attention to whether or not rolling the marbles on a rough surface is different from rolling them on a smooth surface.

Small and Large Marbles Rolling on Rough Surface

(Answers may vary.)

The marble rolled across the carpeted floor from the desk to the chair. It did not roll as far as before. The marble just stopped on the carpet and did not hit anything. It was harder to roll the marble.

Have the students draw what they observed.

II. Think About It

❶ Review with the students Chapter 4 of the *Student Textbook* which covers mass, inertia, and friction. Help the students understand that friction is a force that stops moving objects. Have them apply this idea to the experiment they just performed to show which surface has more friction. Also help the students connect the concepts of mass and inertia—the more mass, the more inertia.

Help the students think about their experiment and any observations they made about the two marbles rolling. You can use the following questions to guide the discussion:

- *Based on your observations, was one marble easier to roll on the smooth surface?*
- *Based on your observations, was one marble easier to roll on the rough surface?*
- *Did one marble go farther than the other marble on the smooth surface?*

- *Did one marble go farther than the other marble on the rough surface?*
- *Once the marbles were moving, did one marble move faster or slower than the other marble?*

❷ Help the students decide which surface has more friction. If the students are not sure, review their results with them. Show them that the marbles had much more difficulty rolling on the rough surface than on the smooth surface and that this tells the students that the rough surface has more friction. They will place a circle around **Rough Surface**.

❸ Help the students decide which marble has the most mass. If the students are unsure, have them hold one marble in each hand. Can they feel which marble is heavier? The heavier marble has the most mass. The larger marble will be heavier than the smaller marble, so it has more mass. They will place a box around **Large Marble**.

❹ Help the students decide which marble has the most inertia. The marble with the most mass has the most inertia. Since the large marble has the most mass, it also has the most inertia. They will place a triangle around **Large Marble**.

III. What Did You Discover?

The questions can be answered verbally or in writing, depending on the writing ability of the student. With these questions help the students think about their observations. There are no "right" answers to these questions, and it is important for the students to write or discuss what they actually observed. Help them explore how the answer they got may be different from what they thought might happen.

Help the students compare the two marbles, and help them explore any similarities or differences between how the two marbles rolled on the different surfaces. They should have discovered that the larger marble is slightly harder to start rolling than the smaller marble, but that once it is going, it is harder to stop (it rolls farther, or takes more force to stop it from rolling). In their discussions, help them use the words "friction," "inertia," and "mass." For example, they might say that the small marble has less mass and less inertia than the large marble.

IV. Why?

The students compared two marbles of different sizes and how they roll on two different surfaces. They may have observed the larger marble rolling farther on the rough surface than the smaller marble. They may also have observed that the larger marble can roll farther than the small marble on the smooth surface. However, it may have taken a bigger push to get the larger marble to roll. Explain to them that what they observed is related to mass, inertia, and friction. The larger marble has more mass and more inertia, and so in order to stop, it requires more friction than the smaller marble requires. Also explain to the students that because of its inertia, a marble in outer space could keep moving without ever stopping because there is no air friction in outer space to stop it.

V. Just For Fun

Have the students observe what happens when a small marble hits a large marble and when a large marble hits a small marble. Ask them to observe if there is any difference, and if so, why. Have the students record their observations.

Experiment 5

Lemon Energy

Materials needed

- 3 large lemons
- knife
- 3 copper pennies older than 1982
- 3 galvanized (zinc coated) nails
- LED (Radio Shack #276-30700)
- 4 pairs alligator clips
- plastic coated copper wire
- wire clippers
- small Phillips screwdriver

[duct tape can be substituted for alligator clips]

Objectives

In this experiment students will explore the concept of stored chemical energy by making a battery from lemons.

The objectives of this lesson are:

- *To have students observe how the chemical stored energy in a lemon can power a small LED.*
- *To help students summarize their observations.*

Experiment

I. Observe It

Assembling the Three-Lemon Battery Electric Circuit

❶ Help the students make slits in each lemon to insert a penny and a galvanized (zinc coated) nail. Make sure the slits go past the lemon rind and into the fleshy part of the lemon. Have the students place a penny in one slit and a zinc nail in the other slit of each lemon. These are the battery "leads."

❷ Using the wire clippers, cut a piece of coated copper wire 15-20 cm (6-8 inches) in length, and strip off about 1.5 cm (one-half inch) of plastic coating from each end. With the Phillips screwdriver, attach a red alligator clip to one end of the wire and a black alligator clip to the other end. Now connect the red alligator clip to the penny of one lemon (the positive lead) and the black alligator clip to the zinc nail of a second lemon (the negative lead). Repeat, connecting the third lemon. Alternatively, you can use duct tape in place of the alligator clips.

You should now have a series of lemons connected by wires with alternating penny and zinc leads. The ends of the two lemons on the outside of the series will have one end of a wire attached and one end free.

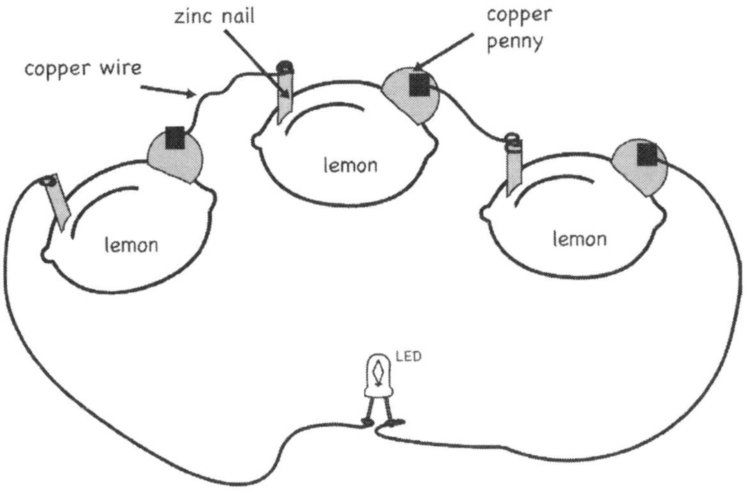

❸ The LED will be connected to the remaining unattached alligator clips or wire ends.

❹ When the students connect the LED to the free alligator clips or wires, the LED should light up. The light may be weak, so you may have to dim the room lights. The LED also has positive and negative leads, so if it does not light up, try switching the wire connections. Also, check the connections for all the pennies and zinc nails. Make sure that metal is touching metal and also that the pennies and zinc nails are deep enough in the lemon to contact the fleshy part of the fruit.

Use the following questions to help the students make their observations.

- *Can you see any light coming from the LED?*
- *Are all the wires securely attached?*
- *Are the pennies and zinc nails in the soft part of the lemon?*

❺ Have the students record their observations in the space provided. They can write a description or draw a picture.

❻ Have the students remove one of the wires attached to one of the pennies. Guide their observations with the following questions:

- *Can you see any light coming from the LED?*
- *Reattach the wire. Does the light come back on?*
- *Remove the wire again. Does the light go off?*

❼ Have the students record their observations in the space provided. They can write a description or draw a picture.

❽-❾ Have the students repeat steps ❻-❼, now disconnecting a wire attached to a zinc nail. Use the same questions as in Step ❻.

Have the students summarize their observations in the chart.

Summarize Your Observations

Trial	What Happened?
All wires connected	*The LED was illuminated*
Penny wire disconnected	*The LED was not illuminated*
Zinc wire disconnected	*The LED was not illuminated*

II. Think About It

❶ Help the students think about their experiment and make any observations about the lemon battery circuit that seem important to them. Use the following questions to help them think about the experiment.

- *How easy or difficult was it to set up the lemon battery circuit?*
- *How bright was the LED when all the wires were connected?*
- *How did the LED respond when you removed one of the copper wires from either the penny or the zinc nail?*

❷ Have the students review Chapter 5 of the *Student Textbook* which covers batteries and chemical energy. Discuss the concept that chemical reactions can produce electrical energy inside batteries. Help them relate this fact to the lemon battery.

❸ Help the students decide which statement is true. They should circle the statement "The LED will light up only when all the wires are connected."

❹ Have the students discuss any problems they may have encountered while doing the experiment. Possible problems are:

- *Dull pennies (if the pennies are dull, there is an oxide layer on the outside of the metal which will prevent electrical contact).*
- *Lemon rinds too thick or too thin.*
- *Wires falling off.*
- *Tape not sticking.*
- *LED is defective.*

Have the students discuss how problems occur while doing experiments and that this is normal in science. Have them discuss possible ways to fix the problems they encountered.

III. What Did You Discover?

The questions can be answered verbally or in writing, depending on the writing ability of the student. With these questions, help the students think about their observations. There are no "right" answers to these questions, and it is important for the students to write or discuss what they actually observed. Help them explore how the answer they got may be different from what they thought might happen.

Help the students compare what happens to the LED when all the wires are attached and what happens to the LED when one or more of the wires are disconnected.

IV. Why?

Have a discussion with the students about how they were able to use lemons as a battery. Also discuss with them how they created an electric circuit by connecting the batteries together. In order to light an LED, at least three lemons are needed. Each lemon generates about 0.5 to 0.75 V of electric current, and an LED generally needs at least 2.0 V of electricity to illuminate. By adding the lemons together in a series, the voltage of each lemon is added to the others, and the total amount of electricity is enough to light the LED. Also discuss with the students how, when they disconnected one of the wires, the circuit was "broken," and the flow of electrons could no longer reach the LED.

V. Just For Fun

First, have the students make sure that all connections are secure, and then have them disconnect one of the wires from the LED. Have them hold the disconnected end of the LED wire in one hand and the end of the wire connected to a lemon in the other hand. Their body is now part of the electric circuit, conducting electricity and causing the LED to illuminate.

Experiment 6

Sticky Balloons

Materials needed

- 2–3 rubber balloons
- string or thread, at least 2 meters (6 feet) cut in half
- scissors
- different materials such as:
  cotton clothing
  silk clothing
  wool clothing
  wooden surface
  plaster wall
  metal surface
  leather surface

In this experiment students will explore static electricity and how charges can transfer from one object to another.

The objectives of this lesson are:

- *To help students observe how an object can become charged.*
- *To help students observe how a charged object can generate an attractive force.*

I. Observe It

In this section students will perform a simple experiment to explore the transfer of static electric charges. They will explore the concept that different kinds of materials and surfaces will donate electrons to a rubber balloon. Rubber has a greater attraction for electrons than some other materials, such as wool, hair, silk, or fur and so will become charged when rubbed against these materials.

❶ Have the students blow up a rubber balloon. They will need to securely tie the end. Have them place the balloon on a wall. The balloon should not be charged, and unless it has picked up charges from being handled, it will simply fall off the wall. This step is called a "control."

❷ Have the students rub the balloon in their hair.

❸ Have them pull the balloon away from their hair and observe whether the balloon pulls their hair. If the balloon pulls their hair, it has attracted electrons from the hair and can be tested for an electric charge. If the balloon does not pull the hair, have the students rub the balloon in their hair again. In humid climates this experiment may not work very well since the moisture in the air prevents the charges from sticking to the balloon.

❹ Have the students "test" for the electric charge on the balloon by placing it on a wall.

❺ Help the students record their observations in the space provided. Use the following questions (also listed in the *Laboratory Workbook*) to guide their inquiry.

- *Does the balloon stick?*
- *How long does the balloon stick? 1 second? 2 seconds? 10 seconds? Longer than a minute?*
- *Does the balloon move around or stay still?*
- *What happens if you blow gently on the balloon? Does it stay stuck or does it fall off? (The balloon should stick to the wall if there have been enough charges transferred from the hair to the balloon.)*

Hair

(Answers may vary.)

The balloon got really charged with my hair. It pulled my hair to the sides. The balloon really stuck to the wall, and even when I blew on the balloon it didn't move. It stayed on the wall longer than a minute.

Have the students draw what they observed.

❻ Have the students repeat Steps ❷-❺ using other materials. To discharge the balloon, wipe the outside of the balloon with a moist paper towel. Once the balloon is discharged, have the students rub the balloon against a different material or surface to recharge it. For each material have the students test the balloon to see how well it sticks to the wall and have them note how long the balloon stays on the wall each time.

II. Think About It

❶ Help the students think about their experiment and make any observations about the different materials they used to create charge on the balloon. You can use the following questions to guide the discussion:

- *Was it easy to get the balloon to carry a charge?*
- *What problems did you have in getting the balloon charged?*
- *Did some materials work better than other materials?*
- *How well did the balloon stick to the wall?*
- *Do you think using a wall to stick the balloon to is a good way to test for charge?*

❷ Review with the students Chapter 6 of the *Student Textbook* which covers electrons, charges, and force. Help the students understand that an electron carries a charge and that what they observed with the balloon is that the charges can jump from one object (hair, silk cloth, fur) to another object (the balloon).

❸ Help the students create a chart listing the materials or surfaces from **Most Charge** to **Least Charge**. The wall test is used to determine how much charge the balloon collected from the various materials or surfaces. This is a qualitative estimation and there may be some surfaces or materials that do not differ much from each other.

You can help the students guess which materials may have had more charge by the length of time the balloon stayed on the wall. The longer the balloon stuck to the wall, the more charges it picked up from the material.

III. What Did You Discover?

The questions can be answered verbally or in writing, depending on the writing ability of the student. With these questions, help the students think about their observations. There are no "right" answers to these questions, and it is important for the students to write or discuss what they actually observed. Help them explore how the answer they got may be different from what they thought might happen.

Help the students explore how easy or difficult this experiment was to perform. Also, help them think about why they may get different results from the different materials used. Some materials may donate more charge to the balloon than other materials. The wall test is a way to determine how "much" charge the balloon collected. Help the students explore whether this is a good way to determine charge on the balloon. Ask the students if they can think of any other way to test the charges.

IV. Why?

Have a discussion about rubber being a material that easily picks up electrons from other surfaces. Discuss the "control" that was performed at the beginning of the experiment. By blowing up the balloon and applying it to the wall before they rubbed it in their hair, they got a feel for how the balloon behaves when it is uncharged.

Depending on the weather conditions in your hometown, this experiment was either easy to perform or did not work well. If the air is very dry, like in arid parts of the country, then static electric charges are easily created. If the air is more humid, then static charges are more difficult to create. Therefore, the results may vary depending on where you live.

V. Just For Fun

Have the students take two balloons and tie a piece of string or thread on one end of each balloon and then tie the other ends of the two pieces of string together.

Have the students drape the tied balloons over a shower rod, or help them fix the balloons to a doorway. Since the balloons are not charged, they should float free, touching each other but not touching any other surface.

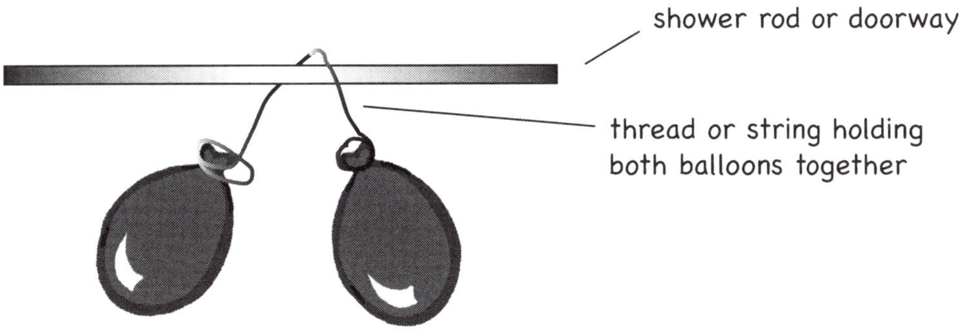

Help the students take the balloons and rub them in their hair. Have them gently let go of the balloons and watch what happens. If both balloons are charged they will repel each other. Explain to the students that when things have the same kind of charge, they repel each other. When they have opposite charges, they attract each other.

Experiment 7

Moving Electrons

Materials needed

- lemon battery supplies (see Experiment 5)
- suggested test materials
  Styrofoam
  plastic block
  cotton ball
  nickel coin
  metal paper clip
  plastic paper clip
- glass of water
- table salt, 15 ml (1 Tbsp.)

Objectives

In this experiment students will explore moving electric current and observe how insulating materials resist the flow of electrons through them.

The objectives of this lesson are:

- *To help students observe the effect of insulating materials on electric flow.*
- *To have students organize results two different ways.*

Experiment

I. Observe It

In this section students will perform a simple experiment to explore how different materials conduct electricity. They will test both metals and nonmetals and compare their results.

❶ Help the students assemble the lemon battery from Experiment 5. Complete the setup exactly as in Experiment 5, and confirm that the LED is illuminated.

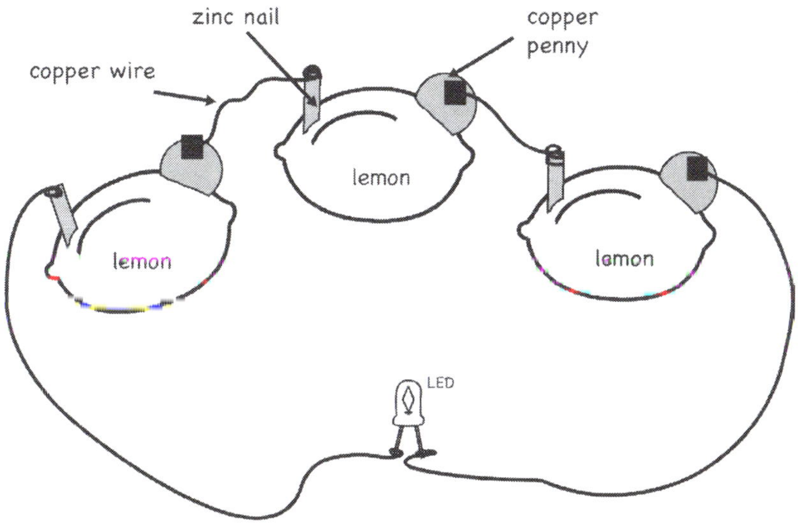

❷ Have the students cut one of the wires that connects two of the lemons. Tell them to observe the LED as they cut the wire.

❸ To reconnect the wire, help the students strip off about 1/2 inch of plastic coating from the ends of the cut wire. Have them reconnect the metal ends by briefly touching them together, or they may want to twist the ends around each other. Make sure that metal is connected to metal and that a metal end has not been wrapped around the plastic coating. Have the students observe the LED and write or draw their observations in the space provided.

❹ Have the students disconnect the cut ends of the wire once again. Have them stick each metal end in a piece of Styrofoam. The ends should not touch each other, and the piece of Styrofoam will complete the connection. Have the students observe the LED and write or draw their observations.

❺ Have the students remove the Styrofoam and reconnect the wires as they did in Step ❸. Have them observe what happens to the LED and then write or draw their observations in the space provided.

❻ Have the students repeat steps ❹-❺ using the other materials on the list. If some of the materials are not available, they can be exchanged for materials that are available. Have the students test a few insulators (plastics, cotton, wool, etc.) and a few conductors (metals).

❼ Have the students summarize their results in the table provided. If an item has been exchanged, have them mark out the listed item and write in the new item. There should be a mixture of LED "ON" and LED "OFF" results as shown in the chart.

Item	LCD
Start—wires connected	*ON*
Wires apart	*OFF*
Wires connected	*ON*
Styrofoam	*OFF*
Wires connected	*ON*
Plastic block	*OFF*
Cotton Ball	*OFF*
Nickel coin	*ON*
Metal paper clip	*ON*
Plastic paper clip	*OFF*

II. Think About It

❶ Help the students think about their experiment and make any additional observations about the different materials they used to test the flow of electrons. Use the questions below to guide their inquiry.

- *Did you notice a difference in how the LED responded to the use of different materials as part of the connection?*
- *What problems did you have connecting the wires to the different materials?*
- *Do you think some of the problems you had affected your results?*
- *Do you think the LED would light up if you used bigger (or smaller) pieces of plastic, Styrofoam, or cotton?*
- *Do you think the LED would be brighter if you used larger pieces of metal?*

❷ Review with the students Chapter 7 of the *Student Textbook* which covers moving charges. Help the students understand that in order for electrons to flow through a material, the electrons have to be free to jump from one atom to the next. Have them think about the fact that insulators are insulators all the way down to their atoms and that using a larger piece of plastic or Styrofoam will not make an insulator conduct electricity.

❸ Have the students organize the materials in the chart provided. This exercise allows the students to explore alternative ways of examining scientific data. Explain to the students that scientists will often reorganize their data in ways that help them understand different aspects of the experiment. Here the students will see that those items that illuminated the LED are grouped together as conductors and those items that did not illuminate the LED are grouped together as insulators.

III. What Did You Discover?

The questions can be answered verbally or in writing, depending on the writing ability of the student. With these questions, help the students think about their observations. There are no "right" answers to these questions, and it is important for the students to write or discuss what they actually observed. Help them explore how the answer they got may be different from what they thought might happen.

Help the students examine the types of materials that illuminated or did not illuminate the LED. It will be clear from the data that plastics, Styrofoam, and cotton are like each other in that they are all materials that do not allow the flow of electrons (i.e., are insulators). It will also be clear from the data that coins, metal paper clips, and other metals do allow the flow of electrons (i.e., are conductors). Help the students use the language of "conductors" and "insulators" to formulate their answers.

IV. Why?

In this experiment students explored how conductors and insulators behave differently. Conductors, such as metals, allow electrons to flow and permit the illumination of the LED. Insulators do not allow electrons to flow and do not permit the illumination of the LED. Explain that the experiment they set up is a good way to determine which materials conduct electricity and which materials do not. Also explain that there are some materials that allow a restricted electron flow. These materials are called resistors.

V. Just For Fun

Have the students take the ends of the cut wire and place them in a glass of water. Depending on the type of tap water you are using, the LED may or may not illuminate.

Have the students add 15 ml (1 Tbsp.) of table salt to the water and stir until it is completely dissolved. The LED should now illuminate.

Discuss with the students how this experiment showed that water and salt water are conductors because they allow electrons to flow through them.

Experiment 8

Magnet Poles

Materials needed

- two bar magnets with the poles labeled "N" and "S"

Objectives

In this experiment students will explore the nature of magnetic poles.

The objectives of this lesson are:

- *To help students observe how magnetic poles behave when placed close together.*
- *To have students make a scientific conclusion based on their observations.*

Experiment

I. Observe It

In this section the students will perform a simple experiment to explore how magnetic poles attract and repulse each other.

❶ Have the students place the bar magnets on a table with the "N" on one magnet facing the "N" on the other. Help the students separate the magnets so that they are far enough apart that there are no repulsive forces interacting.

❷ Have the students gently push the end of one magnet closer to the end of the other magnet. Help them go slowly so that they can observe the stationary magnet. At some point the stationary magnet will change. It will move away from the magnet that is being pushed (the stationary magnet will be repulsed). At this point don't discuss the "North" and "South" poles of the magnets, just focus the students' attention on observing what is happening to the magnets. Help them write or draw their observations in the space provided. This is "Trial 1."

❸ Again have the students place the two magnets on the table several inches apart. Have them reverse the magnetic poles. They can simply flip one bar magnet over so the "S" pole of this magnet is facing the "N" pole of the other magnet.

❹ Have the students gently push the magnets together as they did in Step ❷. This time the magnets should be attracted to each other and may even snap together. This is "Trial 2."

❺ Have the students repeat Steps ❶-❹ several times.

Help the students make observations for each trial. Guide their inquiry with the following questions:

- *How close can you bring the magnets together before something changes?*
- *What happens when you bring the magnets together slowly?*
- *What happens when you bring the magnets together quickly?*
- *Do the magnets behave the same way each time you bring two "Ns" together?*
- *Do the magnets behave the same way each time you bring an "N" and an "S" together?*

❻ Help the students summarize their results in the chart provided.

Trial	N-N or N-S	Together or Apart
Trial 1	*N-N*	*Apart*
Trial 2	*N-S*	*Together*
Trial 3	*N-N*	*Apart*
Trial 4	*N-S*	*Together*
Trial 5	*N-N*	*Apart*
Trial 6	*N-S*	*Together*
Trial 7	*N-N*	*Apart*
Trial 8	*N-S*	*Together*
Trial 9	*N-N*	*Apart*
Trial 10	*N-S*	*Together*

II. Think About It

❶ Help the students think about the magnets and the two different ways the magnets were positioned as they were pushed together.

❷ Review with the students Chapter 8 of the *Student Textbook* which covers magnetic poles. Have a discussion about magnets having two opposite poles that are called "North" and "South." Have the students look at the magnets and observe the "N" and "S" letters.

❸-❹ Have the students look at the results they gathered in the previous table. In this table they recorded N-N or N-S notations along with the observations they made of whether the magnets moved together or apart. Have the students cover up the column that has the N-N and N-S notations, and have them look only at the column that says whether or not the magnets moved together or apart. Using this information, have the students guess which trials had the "same" poles and which trials had the "opposite" poles, and have them record their answers in the table provided.

Trial	Same or Opposite
Trial 1	*Same*
Trial 2	*Opposite*
Trial 3	*Same*
Trial 4	*Opposite*
Trial 5	*Same*
Trial 6	*Opposite*
Trial 7	*Same*
Trial 8	*Opposite*
Trial 9	*Same*
Trial 10	*Opposite*

Next, have the students uncover the hidden column and compare the two tables. Have them note whether or not their answers in the second table match the second column of the previous table.

In this exercise they are using their scientific results to make a guess, predicting which poles were the same and which were opposite based only on their observations.

III. What Did You Discover?

The questions can be answered verbally or in writing, depending on the writing ability of the student. With these questions, help the students think about their observations. There are no "right" answers to these questions, and it is important for the students to write or discuss what they actually observed. Help them explore how the answer they got may be different from what they thought might happen.

Help the students summarize their results verbally or in writing. Have them discuss what happened when they pushed the two "N" poles together and when they pushed the "N" and "S" poles together. Ask them if, based on this experiment, they could take two magnets that were not labeled "N" and "S" and predict which poles were the same and which were opposite. Help them understand that they do not need to know "N" and "S" but only "same" and "opposite." Remind them that opposite poles attract and similar poles repulse and that they can predict which poles are the same or opposite by whether or not the magnets are attracting each other or repulsing each other.

IV. Why?

In this experiment the students explored the magnetic poles of magnets. These poles interact and produce either an attractive or repulsive force. The students took two magnets and, by bringing the ends of the magnets together, they explored how the poles interact with each other.

The students also did several "trials" to explore the magnetic poles. By repeating the steps of the experiment several times, the students could "play" with how the magnets interact with each other, making observations for each trial. Scientists generally do more than one trial when they are conducting experiments. By playing with the experiments, scientists discover things that they might have missed if they only did the experiment once.

V. Just For Fun

Have the students test the magnets with different household items. Have them touch a magnet to a number of different objects, both metal and nonmetal, to find out which materials will interact with the magnet.

Have them record their results in the table provided.

Surface	Attract or Not Attract

Experiment 9

Splitting Light

Materials needed

- 2 plastic or glass prisms
- flashlight
- colored pencils or crayons
- cardboard box (optional)

Suggested

- colored light bulb
- halogen light bulb
- fluorescent light bulb
- mercury light bulb

In this experiment students will explore how a prism can split white light into the different colors of the rainbow.

The objectives of this lesson are:

- *To have students use a scientific tool for examining nature.*
- *To have students compare different results by using a tool in different ways.*

I. Observe It

In this section students will perform a simple experiment to explore how a prism can split the "white light" that comes from sunlight or a flashlight into different wavelengths.

❶ Have the students take one of the prisms and hold it up to the sunlight outside.

❷ Help them rotate the prism until they can see a rainbow reflected somewhere on the ground. The rainbow may be easier to see if the sunlight is reflected on a smooth wall or into a cardboard box (see Step 8).

Sunlight being reflected through a single prism

❸-❹ Have the students use the space provided to draw what they see. They should be able to see different colors of the rainbow. Depending on how the students arranged the prism, they should be able to see—from right to left—red, orange, yellow, green, blue, and violet light, in that sequence. Have the students carefully color the various bands of light they see coming through the prism, and help them notice the sequence of colors and how they are arranged.

Sunlight

(Drawings may vary.)

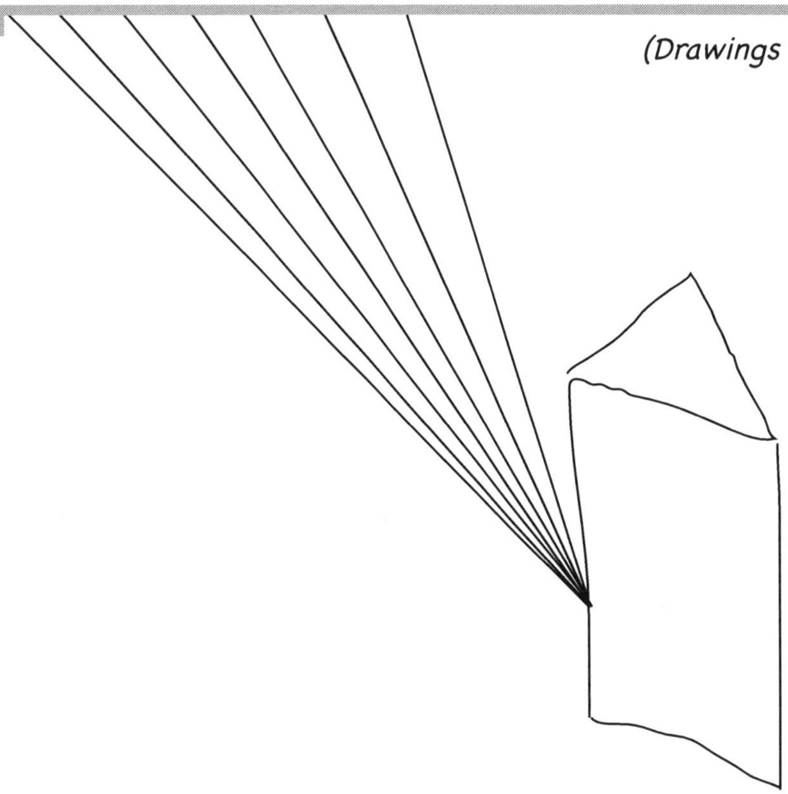

❺-❼ Have the students repeat Steps ❶-❹ using a flashlight for the light source.

❽ ❾ Have the students repeat Steps ❶-❹ using two prisms in sunlight. They should be able to observe an inverted rainbow. Help the students either hold the prisms side by side or have them place the prisms on the ground as shown below.

❿ Have the students repeat Steps ❶-❹ using two prisms with a flashlight for the light source. Again, the students can either hold the prisms in their hands while an adult shines the light through them, or the students can place the prisms on the ground and shine the light from the flashlight through them.

II. Think About It

❶ Help the students think about their experiment and make any additional observations about using one prism vs. two, or about using a flashlight vs. sunlight. You can guide their inquiry with the following questions.

- *What was the order of the colors using one prism and sunlight?*
- *Was the order of colors the same when you used a flashlight?*
- *What do you think this means?*
- *How did the order of colors change using two prisms?*
- *Did you see the same thing happen with both the flashlight and the sunlight?*

❷ Have the students review Chapter 9 of the *Student Textbook* which covers waves and light. Help the students explore the idea that light behaves as a wave. (Light also behaves as a particle, but that is a different discussion). Have the students think about how "white light" is a combination of all the different colors of visible light. Help them understand that the difference between the different colors is the length and/or frequency of the light wave. Shorter, faster waves are blue or violet; longer, slower waves are red.

❸ Have the students examine their drawings and fill in the table.

Light	Shorter or Longer than	Light
green	*shorter*	orange
blue	*shorter*	yellow
orange	*longer*	blue
red	*longer*	violet
violet	*shorter*	green
yellow	*shorter*	red

III. What Did You Discover?

The questions can be answered verbally or in writing, depending on the writing ability of the student. With these questions, help the students think about their observations. There are no "right" answers to these questions, and it is important for the students to write or discuss what they actually observed. Help them explore how the answer they got may be different from what they thought might happen.

Help the students answer the questions about what happened when they placed a prism outside in the sunlight and then in front of a flashlight. Have them compare the results with the flashlight and sunlight. They should have observed similar results.

Have the students answer the question about what happened to the rainbow when they placed two prisms side by side either in the sunlight or in front of a flashlight. They should have observed that the colors became inverted—that is, they changed from violet-blue-green-yellow-orange-red to red-orange-yellow-green-blue-violet.

IV. Why?

In this experiment the students explored the concept that a prism can separate "white light" into the various colors of the rainbow. Explain to the students that "white light" is really a mixture of different colors of light, with each color of light having a different wavelength. The prism bends the light, and when it is bent, the light comes out in different colors because the different wavelengths have been separated.

When the students used two prisms, they should have observed an "inverted" rainbow. The second prism bends the light again, and the effect is to flip the order of the colors. Help the students understand that only the order of the colors is inverted—the sequence stays the same.

V. Just For Fun

In this experiment students can test different types of light sources with the prisms. A fluorescent light bulb, a colored light bulb, a mercury light bulb, and a halogen bulb are recommended.

Have the students record their observations in the chart provided.

Light Source	**Observations**

Experiment 10

Playing With Physics

Materials needed

- 2 marbles
- 3 playing cards
- shallow jar top
- vinegar
- baking soda
- measuring spoons

Suggested

- dominoes
- blocks
- electric car
- electric train
- marshmallow
- tongue depressor
- steel ball

Objectives

In this experiment students will set up a series of small experiments to explore how energy is converted from one form to another.

The objectives of this lesson are:

- *To have students combine several experiments into a series.*
- *To have students observe how energy is converted from one form to another form.*

I. Observe It

In this section the students will use two marbles to design a simple experiment to explore kinetic energy.

❶ Have the students place one marble on the floor. Have them sit some distance away from the stationary marble and roll a second marble into it.

❷ Help the students draw what happens to both marbles. If the marbles are of equal size, the first marble will likely stop, and the second marble will begin to roll when the first marble hits it. The kinetic energy of the first marble is converted into kinetic energy in the second marble.

❸ Have the students arrange three playing cards to make a small card house. The students may need some help in stacking the cards.

❹ Have the students roll a marble, hitting the card house. Have them draw what happens to both the card house and the marble. The cards should fall down, and the marble will likely stop.

❺ Have the students fill a shallow jar top (such as one from a pickle jar) with vinegar.

❻-❼ Have the students add 15 ml (1 tablespoon) of baking soda to the vinegar and have them draw what happens. They should observe bubbles being released from the chemical reaction.

❽-❾ Have the students rinse out the jar top and refill it with vinegar.

Then have them make a card house above the jar top and place 15 ml (1 tablespoon) of baking soda on top of the card house. Next, they will tip the card house with their fingers until the baking soda falls into the vinegar. Have them record their observations.

⑩ Now have the students assemble all of these steps into a short series. Their setup should look something like this:

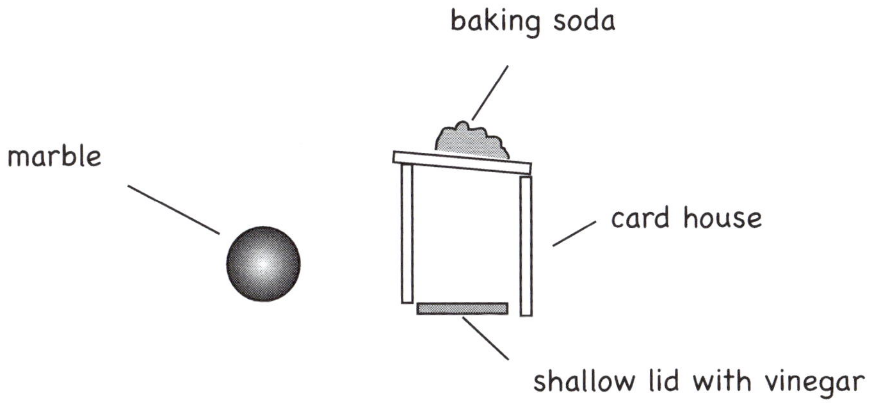

Have the students roll a second marble into the marble that is close to the card house. The marble near the card house should knock it down, causing the baking soda to fall into the shallow lid that contains the vinegar, resulting in a chemical reaction. Have them record their observations.

II. Think About It

❶ Help the students think about the different types of energy they used in this short series of smaller experiments. You can help them think about energy with the following questions.

- *Kinetic energy is the energy of something moving. What was moving in the experiment with two marbles?*
- *What was moving in the second experiment with a card house and a marble?*
- *Did anything move when you added vinegar to baking soda?*
- *What type of energy do you think is used when vinegar is combined with baking soda?*

❷ Review Chapter 10, Section 10.1, *Energy to Energy* in the *Student Textbook*. Have the students think about how energy is converted from one form to another as shown in the illustration given in the book. Explain how energy is never created nor destroyed, just converted from one form to another.

❸ Have the students list the type of energy they think the object started with and the type of energy they think it was converted into. Have them fill out the chart using the energy labels given.

Object	Started With	Converted To
marble-->marble	*kinetic energy (rolling) marble*	*kinetic energy (rolling) marble*
marble-->card house	*kinetic energy (rolling)*	*kinetic energy (falling)*
card house upright--> card house falling	*potential energy (gravitational)*	*kinetic energy (falling)*
baking soda or vinegar -->baking soda + vinegar	*potential energy (chemical)*	*chemical energy*

III. What Did You Discover?

The questions can be answered verbally or in writing, depending on the writing ability of the student. With these questions, help the students think about their observations. There are no "right" answers to these questions, and it is important for the students to write or discuss what they actually observed. Help them explore how the answers they got may be different from what they thought might happen.

Help the students answer the questions about what happened to the energy at each stage of the series of experiments. Have the students think about what might have happened to all the energy at the end of the experiment. Where did it go?

IV. Why?

In this experiment the students explored how energy is converted from one form to another. The students observed how the kinetic energy of one marble is converted into the kinetic energy of another marble. The students also observed how kinetic energy (rolling) was converted into kinetic energy (falling) and how chemical potential energy was converted into chemical energy.

Help the students understand that in each instance energy was converted from one form to another and neither created nor destroyed. Help the students think through the last question. What happened to all the energy at the end of the experiment? There was no more kinetic energy, everything had stopped. There was no more gravitational potential energy, the cards had fallen. There was no more chemical potential energy or chemical energy once the reaction was complete. What happened to the energy? It was converted into a form of energy that is hard to convert—like heat energy. The heat energy was released into the air.

V. Just For Fun

Allow the students to explore the use of different materials to perform energy experiments, connecting them to make a series. Help them make observations about the conversion of energy from one form to another.

Some ideas are provided in the *Laboratory Workbook*.

28809497R00040

Made in the USA
Charleston, SC
22 April 2014